# Editorial Board
## 编委会

| | | | |
|---|---|---|---|
| Chief Consultant | 总顾问 | 余 工 | Yu Gong |
| Consultants | 顾问 | 周晓霖 | Zhou Xiaolin |
| | | 冷 忠 | Leng Zhong |
| | | 余 敏 | Yu Min |
| Editorial Board Director | 编委会主任 | 钱际宏 | Qian Jihong |
| Editorial Board Members | 编委 | 徐 橘 Xu Ju | 李幼群 Li Youqun |
| | | 郭晓华 Guo Xiaohua | 冯华忠 Feng Huazhong |
| | | 刘小平 Liu Xiaoping | 乐永平 Le Yongping |
| | | 冷祖良 Leng Zuliang | 何焱生 He Yansheng |
| Design Director | 设计总监 | 郭晓华 | Guo Xiaohua |
| Chief Editor | 主编 | 郝 峻 | Hao Jun |
| Executive Editor | 执行主编 | 侯江林 | Hou Jianglin |
| Text Editors | 文字编辑 | 北 雁 | Bei Yan |
| | | 梁海珊 | Liang Haishan |
| Art Editor | 美术编辑 | 刘晓蒙 | Liu Xiaomeng |
| Coordinators | 统筹 | 杨 琼 | Yang Qiong |
| | | 冯少芬 | Feng Shaofen |

# Stellar Thoughts

# 星思维

## 第三届"星艺杯"设计大赛获奖作品集

星艺装饰文化传媒中心　编著

暨南大学出版社
JINAN UNIVERSITY PRESS

中国·广州

图书在版编目（CIP）数据

星　思维：第三届"星艺杯"设计大赛获奖作品集 / 星艺装饰文化传媒中心编著. —广州：暨南大学出版社，2015.8
ISBN 978 – 7 – 5668 – 1522 – 4

Ⅰ. ①星…　Ⅱ. ①星…　Ⅲ. ①建筑设计—作品集—中国—现代　Ⅳ. ①TU206

中国版本图书馆 CIP 数据核字（2015）第 151157 号

出版发行：暨南大学出版社

| | |
|---|---|
| 地　　址： | 中国广州暨南大学 |
| 电　　话： | 总编室（8620）85221601 |
| | 营销部（8620）85225284　85228291　85228292（邮购） |
| 传　　真： | （8620）85221583（办公室）　85223774（营销部） |
| 邮　　编： | 510630 |
| 网　　址： | http://www.jnupress.com　http://press.jnu.edu.cn |

| | |
|---|---|
| 排　　版： | 广州良弓广告有限公司 |
| 印　　刷： | 深圳市新联美术印刷有限公司 |

| | |
|---|---|
| 开　　本： | 889mm×1194mm　1/12 |
| 印　　张： | 16 |
| 字　　数： | 190 千 |
| 版　　次： | 2015 年 8 月第 1 版 |
| 印　　次： | 2015 年 8 月第 1 次 |

定　　价：288.00 元

（暨大版图书如有印装质量问题，请与出版社总编室联系调换）

CREATE
HAPPINESS
AND 设计幸福
DELIVER
JOY 播种快乐

# Contents

## 住宅·工程实景作品

| 金奖 | 002 | 2807 号公寓 |

| | 006 | 贵阳山水黔城府邸别墅 |
| 银奖 | 010 | 地标广场 |
| | 014 | 碧桂园凤凰城 |

| | 018 | 博雅首府 A03 公寓 |
| | 022 | 珊瑚天峰 |
| 铜奖 | 026 | 爱的堡垒 |
| | 028 | 山水庭院 |
| | 032 | 潮庭华府 8 栋样品房 |

| | 036 | 北海穗丰金湾样板房 |
| | 040 | 泰城听涛苑 |
| 优秀奖 | 042 | 富川琦园样板房 |
| | 044 | 保利春天别墅 |
| | 046 | 山与城 332 号公寓 |

## 住宅·方案设计作品

| 金奖 | 052 | 星河湾中式别墅样板房 |

| | 056 | 花都玖珑湖别墅样板房 |
| 银奖 | 060 | 星河湾美式别墅样板房 |
| | 064 | 平实人家 |

| | 068 | 林语山庄 20 号 |
| | 074 | 凯旋新世界 T17 |
| 铜奖 | 078 | 保利水晶 |
| | 080 | 君汇世家别墅 |
| | 084 | 汇景新城棕榈园 |

| | 088 | 广西滨江花园别墅 |
| | 092 | 珠江帝景 M8 号 |
| 优秀奖 | 094 | 南沙奥园怡山街别墅 |
| | 096 | 富力泉天下 |
| | 098 | 半岛城邦 |
| | 100 | 广西南宁钻石梦想园 |

## 公共 · 工程实景作品

**金奖** ......... 空缺

**银奖**
- 104 ......... 星艺装饰福建总部办公室
- 106 ......... "墨记"办公室

**铜奖**
- 110 ......... 广西贵港明悦大酒店
- 114 ......... 凤凰博瀚艺术馆
- 116 ......... 830 工作室
- 118 ......... 国宾时光汇售楼部
- 122 ......... 光之游戏办公室

## 公共 · 方案设计作品

**金奖** ......... 126 ......... 北戴河白公馆

**银奖**
- 130 ......... 广西南宁波斯顿精品酒店
- 134 ......... 山水城会所
- 136 ......... 天瑞阳光酒店

**铜奖**
- 140 ......... 车博汇会所
- 144 ......... 桂林香樟林大酒店
- 148 ......... 弘峰国际办公楼大堂
- 152 ......... 愿佛宫殿展厅
- 156 ......... 嘉莉诗国际旗舰店

**优秀奖**
- 158 ......... 清远狮子湖私人会所
- 162 ......... 长征电器办公楼大堂
- 164 ......... 菲音游戏办公室
- 168 ......... 柏川私人会所
- 170 ......... 恒力城奢侈品店
- 172 ......... 南方都市报业传媒办公室

# 住宅・工程实景作品
## Residence · Engineering Live-scene Works

项目名称：2807号公寓　项目设计：广东星艺装饰集团　项目面积：70平方米　项目地址：广东广州　设计师：谭立予　项目用材：黑胡桃实木、白漆

# 2807 号公寓
## Apartment No. 2807

一块原始的水泥板墙面上刻有建筑工人的打油诗，或是莫名其妙的电话号码，以及各种不太完美的印记，设计师果断地把这些元素保留下来。由此一来，这块冰冷的水泥就有了人的情感，并能在未来的生活中继续发酵、升华。

An original cement wall, carved with doggerels by construction workers, random phone numbers, or various imperfect marks, which are decisively maintained by the designer. The cold cements therefore gains human emotions, and will continue to ferment and distill in the future life.

项目名称：贵阳山水黔城府邸别墅　项目设计：广东星艺装饰集团　项目面积：600平方米　项目地址：贵州贵阳　设计师：罗山锐

项目用材：微晶石、实木护墙板、进口墙布、大理石、进口肌理漆、手工油画、艺术PU线、真皮软包

# 贵阳山水黔城府邸别墅
## Landscape Qiancheng Villa in Guiyang

使用者追求奢华空间与贵族的生活方式，注重高雅与富有内涵的生活气质，所以本案设计着重彰显品位、奢华以及体现其独特的视觉效果。设计整体质感对比强烈，线条更是硬朗与柔和并存。

For the owner's pursuit of luxurious living space, noble lifestyle and emphasis on graceful and profound living state, this scenario stresses the presentation of taste, luxury and the unique visual impact generated from the design. The overall design stands out with strong contrasts, with the chorus of softness and hardness in lines.

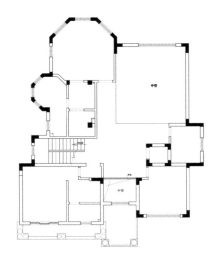
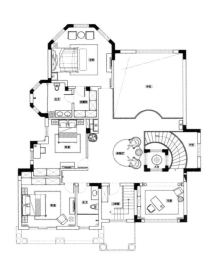

其中客厅的一幅《细水长流》背景画，喷泉浪花顶、玉带盘龙柱楼梯、8字形如意厅以及喇叭状的聚财进口配以元宝入户厅……在每一个细节设计的取"材"及造"形"上都是深思熟虑、精益求精。温馨而不失高贵、华丽而不烦琐，简约而不至于单调。不仅满足了居住者对生活品质与贵气的追求，同时也彰显出内质的生活品位！同时，中国人讲究的是一个好寓意和象征，这便是体现其设计思想的精华所在！

The background picture "Long Flowing Stream" in the living room, the ceiling of fountain spray, the stairs with jade belt and spiral dragon column, the Hall of Auspiciousness in the shape of "8", the trumpet-layout entrance of prosperity gathering and the Entrance Room of Gold Ingot... every detail condenses thoughtful and exquisite material selection and shape styling. Coziness infused with gracefulness, grandness spared from cumbersomeness, simplicity without dullness. The design not only caters for the owner's pursuit of living quality and sense of nobility, but also demonstrates the internal quality of lifestyle! Meanwhile, Chinese highlight the implication and symbol of luck, which is the essence of the design concept.

项目名称：地标广场　项目设计：广东星艺装饰集团　项目面积：106 平方米　项目地址：广东东莞　设计师：吴伟强　项目用材：木纹灰、橡木、水曲柳、灰镜、灰玻、不锈钢、黑镜钢

# 地标广场
*Landmark Square*

限制是设计必要的条件,从使用者的功能需求出发,将有限的空间演化为充满弹性的空间,是设计者在初期思考的重点。用心去探索不同的空间属性,展现独一无二的空间特质,是这个设计案最大的魅力所在。

家,是用来收藏情感和记忆的地方,是涵养心灵的场所,它与坐落位置、面积大小等附加条件无关。"设计"应该以人为主体,依照使用者的实际需求与喜好,搭建人与空间的和谐对话,这是对"家"最好的诠释。

Limitation is the sine qua non of design. Starting from the users' functional requirements and transforming the limited space into flexible domain are the key points in the primary stage of designers' plan. The most attractive charm of this scenario lays in the careful exploration of diversified spatial attributes as well as the presentation of unique spatial features.

Home is the place collecting emotion and memory, the venue of cultivating spirits. It is irrelevant to the extra requirements such as location and area. A "design" should value people as the main part. The best interpretation of "home" rests with the construction of harmonious conversation between people and spaces, according to the users' actual needs and preferences.

# 碧桂园凤凰城
## Country Garden Phoenix City

设计师用精致、大气、纯粹而充满几何感的装饰线条来表现，同时运用明亮的对比色彩来描绘，创造出一种强烈的优雅华美的视觉印象。

Utilizing the sophisticated, splendid and simple decorative lines with geometric patterns while incorporating the bright contrasting colors for depiction, the designer creates strong visual impressions of elegance and magnificence.

项目名称：碧桂园凤凰城　项目设计：广东星艺装饰集团　项目面积：200平方米　项目地址：广东广州　设计师：欧阳乐
项目用材：瓷砖、立邦漆、木地板、不锈钢条

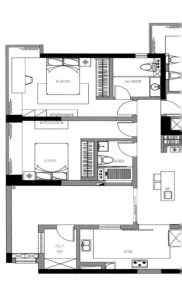

该设计融入了设计师对美学的理解,用简洁的手法完整地展现了优雅的本质,用色彩、光影、物质的不同质感,唤醒生活记忆的润泽与芬芳。这是赠予当今生活在钢筋水泥中的都市人最美好的礼物,让人们体验超越时间与文化界限的生活意境。

设计师在对卧室的处理中,采用了叠加的色调,仿佛用魔力将人拉进空间,墙面材质整体大方。整个空间温馨又不乏时代的摩登感,怀旧的同时又沉浸在自己丰富细腻的情感中,不期而遇的情怀始终闪烁着时光的惊艳。

This design integrates the designer's aesthetic comprehension, comprehensively presents the nature of refinement in plain technique, and evokes the richness and aroma of life memory by different textures of colors, lights, shades and materials. This is the best present for citizens living in the forest of rebar and cements, enabling them to experience the living conception that transcends the border of time and culture.

The designer uses layered colors in the bedroom, which seems to drag people into the space by magic. The texture of walls, natural and poised, paired with the combination of coziness and trendy stylishness—a space for reminiscence, for indulgence in your subtle and sensitive emotions, for the romance of encounters that always sparks the amazement of time.

项目名称：博雅首府 A03 公寓　项目设计：广东星艺装饰集团　项目面积：380 平方米　项目地址：广东广州　设计师：于艳　项目用材：大理石、进口实木地板、杜邦人造石

# 博雅首府 A03 公寓
## A03 Apartment, Erudition Prime Residence

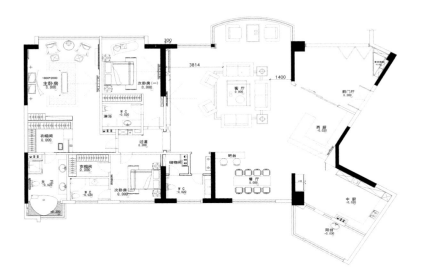

本案入户花园和大厅中超大的过梁都是它的特点，重点在于突出它的美。入户玄关的异形通过划分一个鞋帽间来使其规整，天花上设计的倾斜木顶的结构主要是呼应客厅中间的一根超大过梁。以它为中心制作向两侧倾斜的斜面木质屋面，利用人的视觉落差，抬高整个大厅的层高，同时营造出德式风格特有的木屋感觉。一种集收藏、摆设于一体的墙式书架设计融入其中，功能结构之美，绝对不是为了装饰而装饰。

The indoor garden and the huge lintel standing in the living room outshine in this scenario, aiming to highlight the beauty of the design. The layout of entranceway is orderly adjusted by a locker room, while the slanting wood structure on the ceiling responds to the huge lintel in the living room. With the lintel as the centerline, the oblique wooden ceilings slope on each side, lifting the storey height by visual illusion and generating the sense of a log cabin peculiar to German style. The wall bookshelf, a mixture of collective and decorative functions, seamlessly fits into the design. Such beauty of functional structure, a decoration not for sake.

项目名称：珊瑚天峰　项目设计：广东星艺装饰集团　项目面积：190平方米　项目地址：广东广州　设计师：袁霄　项目用材：地面砖、原木

# 珊瑚天峰
*Coral Sky Cliff*

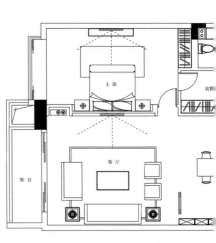

本案以简而不繁的手法解读现代人文的后现代情节。在古典与现代的边缘游走后现代——文脉，是经典而恒久的魅影。轻古典的家装风格摒弃了简约的呆板和单调，也没有了古典风格中的琐碎和严肃，让人感觉庄重和恬静，适度的装饰也使家居空间不乏活泼气息，让人在空间中得到精神和身体上的放松，并且紧跟着时尚的步伐，也满足了现代人的"混搭"乐趣！

在餐厅中以圈椅作为家具，完美地结合了古典与现代设计的元素。将古典与现代相结合，以简洁明快的设计风格为主调，在总体布局方面尽量满足业主生活上的需求。主要装修材料以橡木为主，以及用木栅隔断景点，创造出一个温馨、健康的家庭环境。

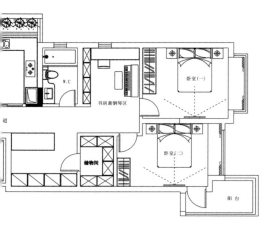

Inviting the technique of conciseness but not complexity, this proposal interprets the post-modernist plots in modern humanity. The nomadic post-modernism wandering by the edge of classics and modernity—cultural lineage, is a classical and lasting phantom. The interior style of light classic tosses the stiffness and tedium in simplicity, erasing the complicatedness and seriousness in classical style. Spraying an air of dignity and tranquility, the proper decoration vitalizes the home space and eases you spiritually and physically. Keeping the pace with fashion, it also satisfies the "mashup" fun of modern people!

Around-back armchairs in the dining room, as furniture, perfectly unite the elements in classical and modern design. An alliance of classics and modernity, the neat and bright design style plays a dominant role, meeting the owner's living requirement in the aspect of overall layout. Oaks are the main finishing materials, with wood fences as partitions, creating a sweet and healthy family environment.

**项目名称：** 爱的堡垒　**项目设计：** 广东星艺装饰集团　**项目面积：** 400平方米　**项目地址：** 江西南昌　**设计师：** 叶惠明

**项目用材：** 大理石、银铂、实木地板、微晶石、无纺墙纸、马赛克、灰镜

# 爱的堡垒
## Fortress of Love

有人说，家是温馨惬意的港湾，因为是她让远航归来的水手有了驻足停留的闲暇；也有人说，家是一杯浓茶，因为是她让经历坎坷的游子品尝到了浓浓的亲情。

家，永远是一个饱含温馨的字眼。本案为新古典主义设计风格，整案采用大面积石材处理，吊顶造型配上华丽的水晶吊灯，展现出奢华、高贵且大气的空间氛围。床头背景墙采用褐色软包，大气且优雅。无论是家具还是配饰均以其优雅、高贵、唯美的姿态，平和而富有内涵的气韵，彰显出居室主人高雅的贵族身份。

Some say home is the warm and comforting harbor, because she offers leisure for the sailors to stop and stay after a long journey; some say home is a cup of thick tea, because she let the children away from home who experience twists and turns taste the rich family affection.

Home, the word leaks warmness forever. This proposal, in the style of neoclassicism, uses wide range of rock materials. The suspended ceiling illuminated by gorgeous crystal chandelier displays luxurious, dignified and grand space atmosphere. Covered by brown soft cases, the background wall behind the headboard radiates the air of greatness and elegance. No matter furniture or decorations, the owner's exalted noble position is always presented in the graceful, noble and beautiful shapes as well as the serene and profound artistic conception.

项目名称：山水庭院　项目设计：广东星艺装饰集团　项目面积：380平方米　项目地址：广东广州　设计师：欧阳乐　项目用材：饰面板、地砖、立邦漆、不锈钢条

# 山水庭院
## *Landscape Garden*

该案例摆脱了金碧辉煌的浮夸装饰风格，重新定义别墅空间，演绎新装饰艺术的奢华感。奢华易造，品位难求。在这一点上，设计师希望通过合理的搭配和恰到好处的装饰来点睛，还原空间原本的优势，从而改造传统意义上的奢华定义，以此来彰显使用者独特的品位。

整体运用素雅、明快的色调，主基调以米色为主，配以亮黑、浅灰以增加色彩层次的对比。黑白灰相间的云石将原本向上集中的空间延伸至四面，增加趣味性，让整个空间氛围形成优雅的奢华感，体现内敛的精细之美。餐厅部分天花的透空设计成为整个空间的亮点，保留童话般的意境，在用餐的同时，时隐时现的星空让人与自然合二为一，和谐共生。

在设计师看来，比"独特"更进一步的是"独到"。客厅部分钻石立体切割造型的白色吧台将人的视觉集中于一点，有收敛中心的作用。透过天花的射灯，突显精妙的清净光源效果。空调出风口下方的深色烤漆玻璃在修饰线条的同时减少了原梁的厚重感，使其更加轻盈，同时深色也和其他部分形成一种和谐共生的呼应感。

Free from the decoration style of pomposity and resplendence, this scenario re-defines the villa space, performing the luxury of new decoration art. Luxury is easy to build while taste is hard to pursue. Starting from this point, the designer attempts to add the striking key point by appropriate matches and proper decorations, restoring the original advantages of space, thereby transforming the traditional definition of luxury and showing the owner's unique taste.

The overall design applies simple, elegant and sprightly colors, with cream color as the base tone, pairing with dark black and light grey for more definition of color contrasts. The marbles, chequered with black, white and grey, extend the space that is originally confined in the upward arch, making the space more enjoyable while forming the graceful luxurious sense and presenting the low-pitched elaboration. The transparent design of the ceiling in the dining hall is the light spot in the space. Maintaining the romance of fairytale, the flickering starry sky seen through the ceiling perfectly unites human and nature, achieving the harmonious intergrowth when you are enjoying the delicacy.

From the perspective of a designer, "unique" is one more step forward than "special". The white bar counter the living room, styled in three-dimensional diamond cutting shape, attracts the attention of every one, effecting as the convergence. The spotlights on the ceiling highlight the delicate and pure light effect. Meanwhile, the dark paint glasses under the outlet of air conditioner embellish the lines and reduce the heaviness and thickness of the beam, lightening the weight in visual impact, responsively coexisting with other parts.

项目名称：潮庭华府 8 栋样品房　项目设计：广东星艺装饰集团　项目面积：150 平方米　项目地址：广东汕头　设计师：余文胜　项目用材：地砖、地板、墙纸

# 潮庭华府 8 栋样品房
## Chaoting Huafu Sample Housing of Building No. 8

一曲琴音一盏茶，且静坐品茗，也品味古典熏陶下的含蓄东方美。本案设计是典型的现代中式风格，中国风的构成主要体现在装饰品和传统家具上（家具多为公司自己设计生产）。

For melodies of Guqin with a cup of tea, just sit still; enjoy the tea as well as the implicit Eastern beauty in classic edification. This scenario employs typical modern Chinese style, which is mainly shown by the decorations and traditional furniture (the furniture is basically manufactured by the company itself).

　　室内的布局采用了对称的布局方式，格调高雅，造型朴素优美，色彩浓厚而成熟。而选用的中国传统室内陈列包括茶具、字画、陶瓷、屏风隔断等，体现了追求者一种修身养性的生活境界。

　　在装饰细节上崇尚自然情趣，花鸟鱼虫富于变化，充分体现出中国传统美学精神。进门对景台与客厅茶几上的摆设给空间增添了中禅的韵味，浓浓的中式韵味在客厅中散发开来。不管是精雕细琢的中式家具，还是精挑细选的软装配饰，都让整个空间更为完美。静茶淡雅、君子淡泊，抿一口清茶，在此体味雅致的人生。

The interior layout employs symmetric structure, displaying graceful style, simple and delicate forms, together with thick and mature colors. The selected traditional Chinese interior decorations include tea set, calligraphy, ink paintings, china, and screens, implying the owner's living state of self-cultivation.

Prioritizing natural interests, the decorative details are full of changes with patterns of flowers, birds, fish and insects, giving rich expressions to the spirit of traditional Chinese aesthetics. The decorations on the view platform facing the door and the tea tables in the living room enrich the space with the appeal of Chinese Zen, radiating Chinese beauty in strong effect. Be it precisely carved and sculpted Chinese furniture, or the careful selected accessories, everything pushes the space closer to perfection. Mild and elegant tea, peaceful and calm gentleman: a sip of the fragrance, a taste of the refined life.

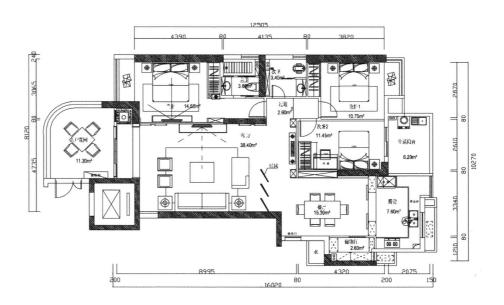

# 北海穗丰金湾样板房

## Sample Housing in Suifeng Gold Bay, Beihai

项目名称：北海穗丰金湾样板房　项目设计：广东星艺装饰集团　项目面积：100平方米
项目地址：广西北海　设计师：许舰 黄林妮　项目用材：黄洞石、雨林棕、泰柚木、绒布硬包

该案例以现代东南亚的装饰手法作为户型的切入点，突出体现悠然的生活情调和闲适的生活气息，逃脱于忙碌的都市生活。材料方面，将黄洞石、雨林棕、泰柚木、绒布硬包这几种肌理材质，运用于室内空间，并通过现代的手法，对东南亚木格元素进一步提炼，营造简练、悠闲、地道的东南亚度假感受。搭配香薰、花瓣、鲜艳色调的点缀，从嗅觉及视觉上给人以冲击。希望能将度假的休闲心情融入生活中，给人们提供一个专属的度假空间。

Adopting the decorative technique of modern South-Eastern Asia as the entry point of house design, this scenario emphasizes the carefree living state and leisurely living climate, escaping from the city life of bustle and hustle. Textural materials including beige travertine, rainforest palm, Thai teakwood and lint hard bags are utilized in the indoor space. Moreover, the South-Eastern wood grid elements are further distilled by modern approach, creating a simple, relaxing and authentic South-East Asia holiday experience. The embellishment of fragrance, petals and bright colors brings about sensational impact into the smell and vision. This scenario aims to provide a customized holiday space, infusing the leisurely holiday mood into life.

# 泰城听涛苑
## Taicheng Tide Music Garden

该设计以简约新中式为主要风格。中式格调多给人沉闷感，为了打破这种感觉，本案在色调上选择比较轻巧明快的颜色，电视背景上的花格造型是以园林中常见的花窗变形而来。整个设计体现了中式的端庄，现代元素的融入又增加了家居的温馨和谐与实用功能。

This design is mainly accomplished in simple neo-Chinese style. Chinese style is often considered to be dull. In order to break such dullness, the scenario selects light and bright colors. The flower lattice background behind the television derives from the lattice window common in Chinese gardens. The overall design not only presents demure Chinese style, but also possesses the warmness, harmony and practical functions of home furnishing with the infusion of modern elements.

项目名称：泰城听涛苑　项目设计：广东星艺装饰集团　项目面积：98平方米　项目地址：天津　设计师：陈伟伟
项目用材：仿木纹砖、麻布硬包、木格、定制墙纸

项目名称：富川瑞园样板房　项目设计：广东星艺装饰集团　项目面积：90平方米　项目地址：广东惠州　设计师：张超宇　项目用材：油漆、墙纸、松木

# 富川瑞园样板房
## Sample Housing of Fuchuan Ruiyuan

本案是精致小三房的样板房设计，考虑到当今年轻人的购房需求偏向于实用性，于是设计中没有太多的豪华奢侈，而是更侧重于温馨舒适的感觉。

设计师摒弃所有多余的为装饰而装饰的造型，将储物空间和设计结合，把房门推拉门和整个过道及沙发背景组合出连贯完整的空间，结合轻松、大方的配饰，力图让每一个看房的客户都能体会到开发商用心规划的90平方米的小三房的超实用性，以及为每一个客户设身处地考虑的温暖感受。

This scenario is the sample home design for an exquisite small apartment with three rooms. Given that the property demand of young people tends to focus on practicality, there are more cozy and sweet elements instead of luxury or expense in the design.

The designer abandons all the redundant decorations that are just for the sake of decorative use. Integrating storage space with design, the sliding doors, the whole hallway and the background behind the sofa unite as a coherent and complete space, embellished with some light and natural accessories. The designer strives to show customers the practicality of the 90m² small apartment that is attentively planned by the developers, as well as the warmness within the thoughtful considerations for all the customers.

项目名称：保利春天别墅　项目设计：广东星艺装饰集团　项目面积：280平方米　项目地址：贵州贵阳　设计师：罗珍
项目用材：仿古砖、实木、墙纸、大理石

# 保利春天别墅
Poly Spring Villa

本案采用东南亚与简中混搭的风格，将中式与异域元素完美融合。整个设计中自然的木纹、简洁的造型、深蓝色的仿古砖都使空间充满自然之趣。

软装搭配上竹叶形的墙纸、藤蔓的窗帘，再加入中式水墨画做点缀，营造出一个具有文化韵味的艺术空间。

客厅墙上的大树造型，巧妙地将自然元素融入空间中，是本设计的一大亮点。空间在满足业主功能要求的同时，并没有掩盖其自然宁静的悠然之美。撩开渐欲迷人眼的乱花迷雾，不加粉饰，回归初心。

This scenario is a mixture of South-Eastern Asia and simple Chinese style, perfectly sewing Chinese and exotic elements into one. The natural wood grains, neat forms and dark blue archaized bricks all energize the space with natural interest.

Soft furnishing joined by wallpapers of bamboo leave patterns and curtains of cirrus design, with the decoration of Chinese ink paintings, an artistic space filled with cultural charm is thereby created.

The big tree on the wall of living room skillfully welcomes natural elements into the space, which is one of the highlights in the design. The space, satisfying the owner's needs, does not cover the peaceful beauty of nature and serenity. Unveil the anthemia and misty fogs that amaze your eyes, erase the makeup and decoration, return to the original soul.

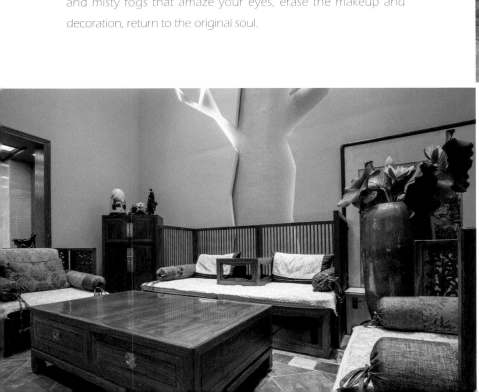

项目名称：山与城 332 号公寓　项目设计：广东星艺装饰集团　项目面积：99 平方米　项目地址：广西桂林　设计师：张抗
项目用材：半抛光砖、复合地板

# 山与城 332 号公寓
## Mountain and City Apartment No.332

如果说家具的款式能够体现空间的气质，那么色彩的搭配就能丰富空间的表情，而灯光则能活跃空间的气质。本案的设计正好印证了这几点。

If the styles of furniture display the spatial temperament, the match of colors will enrich the spatial expressions, and the lights can vitalize the spatial air. This scenario is a perfect verification of the above points.

本案为 99 平方米小三房，因此在进行平面设计时，设计师在对空间和功能的发掘与使用上始终力求完美，大胆改变了原建筑的平面设计，充分利用建筑空间可延展或可利用的位置进行合理设计与再利用。其中客厅阳台设计后的空间涵盖了收纳区、休闲区及工作区三种功能；设计造型上以现代简约的手法融合复古风格的家居及软装饰，色彩搭配上大胆运用色块的强烈对比，丰富了空间的层次感，让空间在复古的优雅中不失青春向上的活力；同时，灯光的设计更是在功能实用的基础上，让空间的明度及纵深度更加丰满。小空间大利用，设计对于生活不仅是一种装饰，更是一种需要，这是设计师对生活的明朗态度，对空间的睿智理解。

This is a design for a 99m² small apartment, therefore the exploitation of space and function together with practicality become the goals of perfection. The designer boldly alters the original design of the architecture, making full use of the extendible and available architecture space for scientific design and reuse. The original living room and balcony are transformed into spaces uniting storage zone, recreation zone and work zone. For styles and forms, the modern simple techniques blended with vintage furnishing and soft decorations, and the intense contrasts in shades and color lumps, add more definition to the spatial layers, maintaining the vintage elegance without the loss of youthful vitality. Meanwhile, on the basis of functional practicality, the design of lights creates more brightness and depth for the space. Small space, big use. Design is not just a decoration for life, but more of a need. This is a designer's positive attitude towards life, and the wise comprehension of space.

# 住宅·方案设计作品
## Residence · Scenario Design Works

项目名称：星河湾中式别墅样板房　项目设计：广东星艺装饰集团　项目面积：260平方米　项目地址：山东济宁　设计师：集团设计研究院
项目用材：橡木、翡翠木纹大理石、玻璃、壁纸

# 星河湾中式别墅样板房
## Chinese-Style Sample in Villa of Star River Bay

狭窄细长的建筑格局是小型联排别墅的常态，这样的格局更加需要充分利用好建筑两头的宝贵采光和可能的通风。设计师兼顾建筑的内外形态结构，最终将客厅区域的主采光大窗做了重新规划。

The narrow and long architecture structure is common in small town houses. This layout stresses the best use of precious lights on the two ends and possible ventilations. Taking both the internal and external forms of the building into consideration, the designer finally re-organizes the big window which is the main lighting source in the living room.

本案将风格定位为简约东方，同时，设计师力求可以用更精练持久的手法传承表达好这一风格，从中式花窗提炼出的"线"最终组合成结合微妙光源效果的大幅的"面"，并一直延展拉伸至整个建筑的主立面，直至贯穿到二、三楼，让原本狭小局促的空间格局显现出宏观的气势。楼梯处于这个"面"的前方，以纯白的质感极简表达，富有张力，舞动中不乏禅韵。

While this scenario positions itself as simple Eastern style, the designer tries best to elaborate and express this characteristic in more concise and lasting approaches. The "lines" distilled from Chinese lattice windows finally constitute a huge "surface" with subtle lighting effect, and extend to the main facade of the whole building, straightly penetrating through the second and the third floor, creating grandness in the small layout. The stairs are at the front of the "surface", painted in simple whiteness, which is highly intense, dancing with the spirit of Zen.

项目名称:花都玖珑湖别墅样板房　项目设计:广东星艺装饰集团　项目面积:1 100平方米　项目地址:广东广州　设计师:集团设计研究院
项目用材:蓝贝露大理石、黑檀木、玻璃、人造石、烤漆板

# 花都玖珑湖别墅样板房
## Sample of Huadu Jiulong Lake Villa

这是一个承载了业主对生活品质和配套的高要求的独栋住宅建筑。原本八百多平方米的体量仍然不足以容纳需要构建在其间的许多生活配套功能。于是设计师对建筑做了巧妙的改动和扩建，并结合各个功能的特性和整个建筑生活人流的可能，做了全新的规划。从容地将酒窖、健身房、SPA、影音室等诸多配套功能区设立其间。花园中的泳池设在整栋建筑围合的中心位置，结合增设的户外观景电梯，相映成趣，人行楼梯设在与泳池同一轴线之上的室内中央区域，以简洁的质感和富有动感的手法塑造，时尚气息自然流露。

This is a house bearing the owner's high requirement of living quality and facilities. The original 800m² was not enough to contain the various living facilities and functions to be constructed. The designer, therefore, subtly changes and re-constructs the architecture and makes brand new plans according to the functional features of different parts and the possibilities of people flow in this building. Different kinds of facilities including cellar, gymnasium, SPA, video room are properly set in the architecture. At the center of the building quadrangle seated the swimming pool of the garden, casting beautiful reflections with the additional outdoor scenic lift. Set on the indoor central area above the axis, the stairs and swimming pool are molded in simple characteristic and mobile techniques, naturally radiating the sense of fashion.

项目名称：星河湾美式别墅样板房　项目设计：广东星艺装饰集团　项目面积：360 平方米　项目地址：山东济宁　设计师：集团设计研究院

项目用材：樱桃木、翡翠木纹大理石、银铂、壁纸

# 星河湾美式别墅样板房
## American-Style Sample in Villa of Star River Bay

This scenario fully respects and keeps the original advantages of the architecture, especially the high slanting roofs on the top storey. Through the reorganization and reassembly of spatial interfaces, the designer creates a vast and high space layout in American residence style, which constructs a wonderful skeleton for implanting special elements and techniques in later stage. The smooth veneer wallboard and functional shelves attached to the walls are the main parts of the design, fitting perfectly into the space. The layered effect of the ceiling breeds vigorous dignified style.

本案设计师充分尊重并保留了建筑原本的优点，尤其是顶层高挑的坡屋顶，并通过空间界面的重新规划组合，营造出一个美式住宅宽大高挑的空间格局，为后期植入应用的专属元素手法提供了良好的骨架。细腻的木质墙板结合墙身一体化的功能柜体作为主基调，协调统一地应用在空间之中。富有层次感的天面处理手法让场景衍生出生动的尊贵型格。

项目名称：平实人家　项目设计：广东星艺装饰集团　项目面积：800 平方米　项目地址：广东广州　设计师：邱崇坚
项目用材：水泥、钢筋、纸叠板

# 平实人家
## Ordinary Family

　　一为文化，二为记忆。该设计旨在以故事告慰那些梦中残存的记忆与期许。"四喜临门""让我们荡起双桨"……设计师试着以自己的情感方式诠释同代人的生活，也试着以一种别样的设计理念与方式来解释人、情感与空间的关联。

　　一个素色可自由处置的建筑空间，一个可让思想与情感自由畅想的地方。尽用一代80后的情感记忆在空间中组织着妙趣横生的设计，其间表达的必是设计者非一般的经历。

For culture, and for memory. This scenario aims to comfort the remaining memory and expectations by stories. "Four Blessings Descend upon the House", "Let's Row the Oars"... The designer attempts to clarify the life of people in the same generation in personal emotions, and to explain the relationship among people, emotion and space in a unique design philosophy.

An architectural space in plain colors, a place free for arrangements, a venue for boundless imagination of thoughts and emotion. Composing interesting designs according to the emotional memories of the generation born in 1980s, the expression must be ralated to unusual experience of the designer.

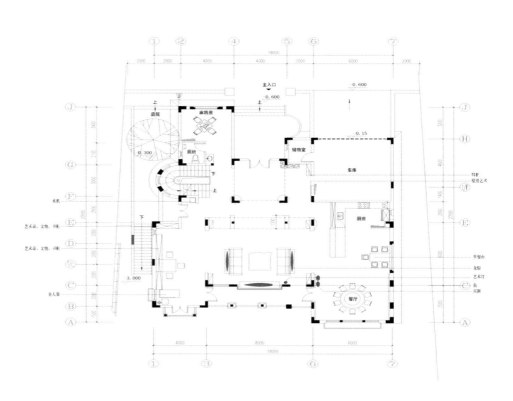

067 - 星 思维

项目名称：林语山庄 20 号　　项目设计：广东星艺装饰集团　　项目面积：1 070 平方米　　项目地址：广东广州　　设计师：集团设计研究院

项目用材：户外：白麻（花岗岩）、中点黑金沙（皇室啡花岗岩）　室内：雅士白、金啡、黑白根大理石、科技橡木、黑色马赛克

# 林语山庄 20 号

*No.20 Whisper of Forest Villa*

该设计根据布局的功能分布，追求视觉效果与实用性的统一，"将最好的风景交给业主""以业主生活为中心"是此次设计的原则，依据功能需求，强调空间形体的同时，在节奏感、私密性、舒适度上都做了细节的处理。高尚会所品质构筑精致空间。

This design is structured according to the layout of functions, pursuing the integration of visual impact and practicality. "Presenting the best view to the owners" and "owners' life is the center" are the principles of this design. Motivated by the functional demands, this scenario conducts detailed dispositions of the rhythm, privacy and comfort degree while paying attention to spatial form. With the quality of top club, construct exquisite space.

项目名称：凯旋新世界 T17　设计施工：广东星艺装饰集团　项目面积：680 平方米　项目地址：广东广州　设计师：姚毅华
项目用材：雅士白大理石、王子玉大理石、艺术涂料、巴西高山花梨等

# 凯旋新世界 T17

## New World of Triumphant Return T17

凯旋新世界位于广州市天河中心区中轴线上,是珠江新城的首席中央居住区,物业南望珠江秀水,北依珠江公园。

本案采用法式装饰风格,其优雅、高贵、浪漫、诗意与情境力求在气质上给人深度的感染。设计中,力求使空间尽情挥洒源远流长的神韵,在给予空间生命力的同时也饱含文化品位。

New World of Triumphant Return is seated on the city central axis. As the prime central residence in Zhujiang New Town, it is to the north of Pearl River and to the south of Pearl River Park.

This scenario uses French decoration styles, whose elegance, refinement, romance, poetic conception and plotting strive for profound temperamental influence on people. The design strains after the sprinkle of the spirit of long standing and the collection of cultural taste with incubation of vitality in the space.

选材方面，除了欧式惯用的大理石材质外，还用了雕花，雕像，装饰画，大大提升了原质感的对比效果，在表现尊贵的同时增添了几分优雅、浪漫情怀。与此同时，本案配搭了暖色的灯光，打破了传统布局沉闷的格调，使得空间的层次感、穿透性大大增强。在色彩运用上以金色和白色为主，以水晶器皿、花艺、装饰品的点缀色为辅，在工艺上雕花线条等制作工艺精细考究。

后期软装方面，选用的配搭灯具、家具等也很考究，既有西式元素，又具有强烈的时代感元素。例如天花水晶灯的闪烁效果，配搭上欧式家具、水晶灯、镜子、挂画、饰品等完美组合，形成极具品位的装饰效果,将奢华的欧式风情演绎得淋漓尽致。

整套方案完美而和谐地融合了空间与人文，突显出居室主人的精神文化品位和审美高度。

Concerning the materials, in addition to the marbles common in European styles, carvings, sculptures and decorative pictures greatly intensify the contrasts of the original textures, adding flashes of elegance and romance to the presentation of nobility. At the same time, the warm light colors break the stiff style in traditional layout, facilitating the layering and sense of depth in the space. The dominance of gold and white is enriched by the supplementary shades of crystal utensils, floristry and decorations. The carved lines are accomplished by sophisticated techniques.

The decorative lights and furniture are selected by careful consideration in soft furnishing of late stage, incorporating both Western elements and strong sense of the present time. For instance, the glow of crystal lights on the ceiling, paired with the perfect combination of European-style furniture, crystal lights, mirrors, wall pictures and accessories, forms a tasteful decorative effect and incisively performs the luxurious European appeal.

This scenario perfectly infuses the harmonious coexistence of space and humanity, stressing the owner's spiritual and cultural state and aesthetic level.

项目名称：保利水晶　项目设计：广东星艺装饰集团　项目面积：700平方米　项目地址：广东广州　设计师：帅伯尤
项目用材：大理石、黑玻、镜钢、玻化砖、软包、进口墙纸

# 保利水晶
*Poly Crystal*

本案是位于广州花都区的别墅，背山靠湖，风景优美怡人。整体设计采用时尚简约风格。

本案中大厅的挑高空间处理，灵动又不失大气，雅致又不失华丽。餐厅的水晶吊灯，卧室柔和的灯光处理给我们带来丝丝细腻的关怀，整个空间软装上的色彩协调到位、奢华大气，实现了湖畔佳苑雍容生活和现代舒适生活的统一。

This scenario is a villa located in Huadu District, Guangzhou. Backed by the mountain, facing the lake, the scenery is graceful and refreshing. The overall design applies fashionable and simply style.

The high space in the hall gives the area vigor while maintaining the grandness, holding the sense of refinement without the lack of magnificence. The crystal droplight in the kitchen and the soft lights in the bedrooms show the sensitive solicitude to the owners. Colors of the soft furnishing are harmonious and luxurious, realizing the unity of noble life by the lake and comfortable modern living state.

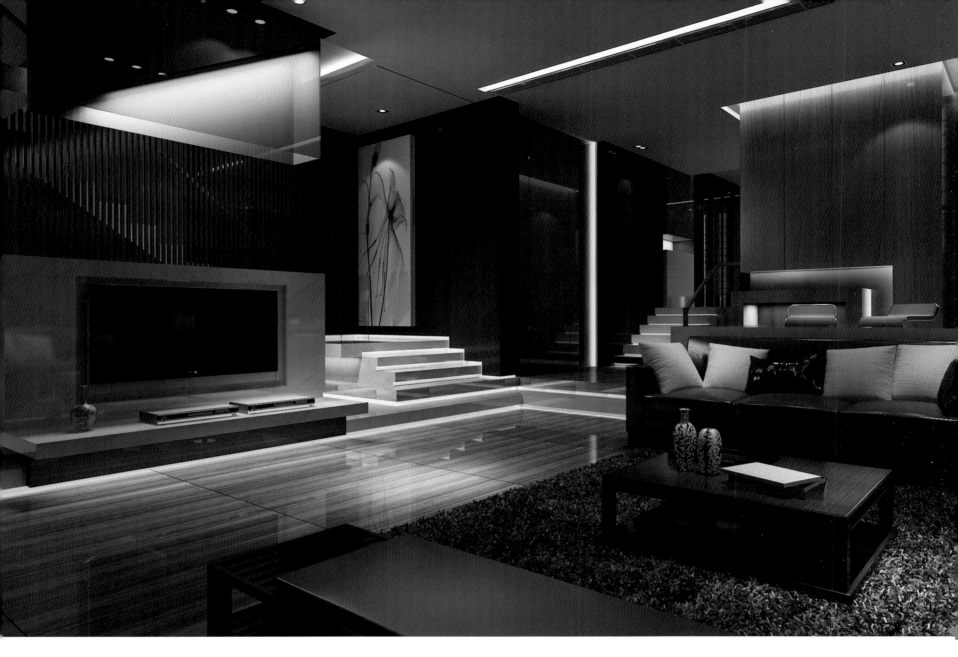

项目名称：君汇世家别墅　项目设计：广东星艺装饰集团　项目面积：500 平方米　项目地址：广东广州　设计师：汪克成　项目用材：玻璃

# 君汇世家别墅
## Junhui Lineage Villa

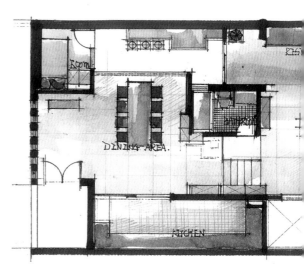

随着生活质量不断提高,人们对赖以生存的环境开始重新考虑,并由此提出了更高层次的要求,特别是对生活水平和文化素质的提高及住宅条件的改善。于是"室内设计"已不再是专业人士的专利,普通百姓参与设计或自己动手布置家居已形成风气,这就给广大设计人员提出了更高的要求。

简约风格是近来比较流行的一种风格,本案设计体现了人们追求时尚与潮流的同时,也非常注重居室空间的布局与使用功能。

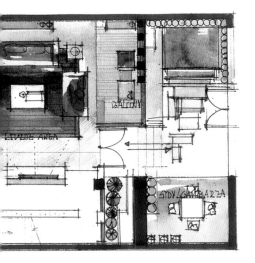

With the advancement of living quality, people start to reconsider the environment that they live in, and put forward high requirements of life, especially the improvement of living standard, cultural quality and housing condition. As a result, "interior design" is no longer a privilege of professionals. It is popular for the common mass to participate in the design and decoration of their homes, which raises the requirement for designers.

Simplicity is quite popular in the recent trend. This scenario prioritizes the fashion and trend and commits the layout and practical functions of spaces.

项目名称：汇景新城棕榈园　项目设计：广东星艺装饰集团　项目面积：280 平方米　项目地址：广东广州　设计师：张苏丽
项目用材：仿古砖、木材、大理石

# 汇景新城棕榈园
Palm Garden of Favorview Palace

禅的精髓就是"不说破"而留给他人思考的余地，相信他们能够思考。通过自己的思考得出结论，比强行灌输进去的效果要好得多，任何说破的东西都是别人的，只有自己悟到的，才是属于自己的。

The essence of Zen sets foot in "not laying bare", leaving enough contemplative space for others, and believing their capacity in thinking. It is much better to draw a conclusion through independent thoughts than being force-fed. Anything laid bare belongs to others. Only those realized on your own belongs to you.

大凡美的东西，都是自然的流露。正如佛语所言："真佛只说家常话。"禅宗的精神便是崇尚自然。设计也应当追求禅宗的家常境界，亲切而自然，平凡而朴素。

All the beauty is natural revelation. As said by Buddha, "The real Buddha only utters ordinary words." The spirit of Zen is to worship the nature. And design, of course, should also pursue the ordinary state of Zen, kind and natural, common and simple.

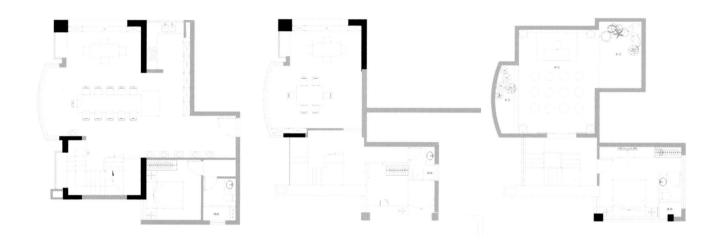

The design concept of ordinary state initiates healthy living philosophy. Though short of material satisfaction, modern people should keep a vigorous spiritual life, returning to ordinary people, ordinary affairs and ordinary life. The design phenomenon of blindly emphasizing luxury, grandness, fuss and gloss-over should be overcome. Despite of the differences between Zen and secular values, it is neither recondite and nihil nor otherworldly. In fact, Zen is a living philosophy, which stresses the expression of "soul" and shows the spiritual performance, so as to pursue tranquil self-reflection and maintain the aloof spiritual state.

家常境界的设计观念是倡导健康的生活观念，即使物质生活清贫，现代人也要保持精神生活充实，力求回归到平常人、平常事、平常的生活中去，克服那些一味强调奢侈的、宏伟的、无病呻吟的、化妆般的设计现象。虽然说禅宗是一种有别于世俗的人生观与价值观，但并不是说它就高深虚无，甚至不食人间烟火。事实上，禅就是一种生活哲学。重视"心"的表达，张扬其精神表现，以从中寻求空寂的内省，保持一种超脱的心灵境界。

项目名称：广西滨江花园别墅　项目设计：广东星艺装饰集团　项目面积：500平方米　项目地址：广西南宁　设计师：陈思成
项目用材：大理石、硅藻泥、马赛克、水泥挂件、墙纸、不锈钢

# 广西滨江花园别墅
## Riverside Garden Villa in Guangxi

本案是对法式贵族复兴的诠释，法式的白色浪漫主义通过空间分割与材料的运用，在细节处体现了空间的唯美与极致奢华，以及主人的高品质生活。从整体到局部，更像一种多元化的思考方式，将古典的浪漫情怀与现代人对生活的需求相结合，兼容华贵典雅与时尚现代。

This scenario is a renaissance of French aristocracy. Through the division of space and utilization of materials, the details of French white romanticism radiate the aesthetics and luxury of the space, together with the high quality life of the owner. From the entirety to the part, a diversified thinking pattern incorporates classical romance into the modern's living requirements, inclusive of grandness, elegance, fashion and modernity.

在立面上使用古典纯白色的线条墙板，搭配着现代纹样的墙布；在地面采用石材拼花，用石材天然的纹理和自然的色彩体现整个空间的品质感，使客厅和卧室的奢华档次和品位毫无保留地流淌。

The white classical striped wallboard on the facade are paired with wall cloth in modern patterns. On the floors are the stone mosaic medallions, whose inartificial textures and natural colors elaborate the quality of the whole space, willfully spraying the luxury and noble taste of the living room and bedrooms.

在家具配置上利用新古典的效果，不但能将木皮的纹理尽情展示，而且在徒手触摸时还能感受到油漆饰面后的平整和光滑。

在软装配饰上使用白色、浅黄色、暗褐色为基调，再加之少量金色柔和，使色彩看起来明亮大方，使整个空间给人以开放、宽容的贵族非凡气度。

The furnishing employs neo-classic effects, which not only thoroughly presents the veneer textures, but also lends a touch of smoothness and neatness of the varnished surface.

White, light yellow and dark brown set the tone for soft furnishing decorations, mixed with a dab of soft gold, brightening the colors and bringing about an open, wide and extraordinary noble temperament of the space.

项目名称：珠江帝景 M8 号　项目设计：广东星艺装饰集团　项目面积：160 平方米　项目地址：广东广州　设计师：陈桂军　项目用材：实木、艺术漆

# 珠江帝景 M8 号
## Regal Riviera No. M8

空间赋予人生的记忆，是"家"应该承载的一部分。

本案位于珠江江畔，设计师以现代简约为设计风格，把质朴与时尚置入空间，创造出低调而优质的自然之家。空间以白色为主色调，富有质感的自然木纹是向往本质的回归，客厅主墙以实木的肌理安排，加强与地面界面关系的连贯整体性，自然且大气。客、餐厅空间的连贯主体面也体现出一种生活互动关系，加大了使用机能，饱含着一种生活温度。

Endowing a space with life memory should be a part carried by "home".

This scenario is located by Pearl River bank. Styling and modern simple design, the designer equips the space with simplicity and fashion, creating a natural home space with low-profile and high quality. White rules the space. The textural natural wood grains lead the way back to the nature. The solid wood on the main wall of living room enhances the coherence and integrity with the floor, which is natural and grand. The coherent main surfaces in living room and dining room also present a kind of life-interactive relationship, broadening the function scope and filling with the warmness of life.

项目名称：南沙奥园怡山街别墅　项目设计：广东星艺装饰集团　项目面积：750平方米　项目地址：广东广州　设计师：于艳
项目用材：玻璃、大理石、瓷砖

# 南沙奥园怡山街别墅
## Yishan Street Villa in Ao Garden, Nansha

该别墅位于风景优美的海边，现代简约的风格，以"光与影"为主题，整体空间以黑白色调为主，显得整个空间简洁、大方，且富有节奏韵律。光与影的完美结合，起到了"画龙点睛"的作用，在天花上，筒灯作为照明的主要光源，它又与暗藏灯相结合，使整个空间在艺术氛围上达到协调统一；光影打在光滑的大理石台面上，所折射出来的艺术效果，正是简约现代设计中突出的重点。

在材质的处理上，本案使用了玻璃、大理石、瓷砖等，材料品种简单但是协调统一。其实材料不在于多，而是在于把不同的材料按照黄金分割的比例穿插在一起。这样既有理性的空间与结构，又有感性的环境与气氛。这正是本设计的核心目标。

Situated by the scenic sea, this villa is designed in modern simple style. Using "light and shadow" as the theme, the whole space is a territory of black and white, which makes it concise, natural and rhythmic. The seamless integration of light and shadow is a finishing touch. The down lights on the ceiling not only function as the main illuminants, but also achieve the artistic conception of harmony and unity with the help of hidden lights; the lights and shadows project on the smooth marble platform, reflecting an artistic impact that is the essence in simple modern design.

Regarding the materials, this scenario adopts glasses, marbles, tiles, etc. Simple yet harmonious as one. It does not matter for the amount of materials, but the interweave of different materials according to the portion of golden cut, realizing the coexistence of rational space and structure as well as emotional environment and atmosphere. This is exactly the goal of this scenario.

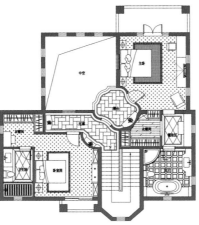

项目名称：富力泉天下　项目设计：广东星艺装饰集团　项目面积：260平方米　项目地址：广东广州　设计师：徐惠惜

项目用材：雅士白大理石、木纹砖、黑板漆、涂料

# 富力泉天下
## Fuli Kingdom of Spring

本案借鉴了米兰流行的色彩,明快的色彩对比使得设计富有现代艺术气息,强烈的补色对比是对传统的挑战和对自由心态的表达;电视背景运用黑板漆,给喜欢绘画的孩子一个无限涂鸦的空间;真皮地毯是不可缺少的元素,这是新贵们的标配,一种低调的张扬,散发着无法掩饰的自然气息,也温和地激发灵感与想象。

主卧的床头背景也大胆运用了色块的对比,大面积的纯色块宁静悠远,而且容易受到不同时间段光线变化的影响,空间的氛围也会随之变化,它仿佛在提醒我们每一刻都是崭新的开始,每一刻都应该享受和珍惜。

Colors popular in Milan are used in this scenario. Lively colors lend some hints of modern art to the design and sharp contrasts of complementary colors represent the challenge to tradition and the heart for freedom. The background of the television is painted with blackboard paint, providing a vast space for graffiti for children who love painting. Leather carpet is indispensible, a must for the new nobles as well as a low-profile expression. It also radiates the air of the nature and mildly activates inspiration and imagination.

A contrast of colors is also boldly applied in the bed background of the master room. Large chunks of colors radiate tranquility and serenity. Subject to light changes in different stages, the atmosphere of the space changes accordingly, as if telling us that every moment can be a brand new starting point, and that every moment should be cherished and enjoyed.

项目名称：半岛城邦　项目设计：广东星艺装饰集团　项目面积：220平方米　项目地址：四川成都　设计师：景晓　项目用材：大理石、木地板、饰面板、墙纸

# 半岛城邦
## Peninsula Polis

本案设计风格围绕"现代简约"这个主题，以线条作为设计元素，以木纹石材及木地板上墙作为点缀，注重居室空间布局，流线简单明快，没有过多的装饰，一切从视觉出发。

This scenario features modernity and simplicity, utilizing lines for design. Stones with wood patterns and wood flooring on walls are used for decoration. The focus is on the interior layout. The lines are simple and lively, without excessive decoration. The visual effect is prioritized in this design.

以深色为主的家具配置搭配极富对比色的简洁饰物，使整个环境格调清新。以灰色调为居室空间的主色调，体现了现代人的生活节奏，以及时尚前卫的生活方式。

Furniture in dark colors goes with decorations rich in complementary colors, which gives refreshing air to the whole space. Grey is the most used color in this design, a symbol of modern people's pace of life as well as their fashionable lifestyles.

项目名称：广西南宁钻石梦想园　项目设计：广东星艺装饰集团　项目面积：38平方米　项目地址：广西南宁　设计师：彭丽萍
项目用材：瓷砖、地砖、饰面板、墙纸、镜面

# 广西南宁钻石梦想园
## Nanning Diamond and Dream Garden in Guangxi

The room is shaped as a cuboid, with the balcony being the only source of natural light, so that wardrobes are used to divide the whole room. Given that the rectangle is not long enough, the length is extended to the balcony to prolong the space and resolve the problem of lighting. For lighting, the designer has pasted diamond pad onto the transparent glass between and on the top of the wardrobes. There are diamond-shaped openings in the partition near the desk for letting in daylight. Through such means, the problem of lighting of the living room and the kitchen could be fully solved.

There is a movable table in the aisle between the kitchen and the living room. Between meals, the table can be moved to the side of the television background for holding decorations. The table is firmly supported and it is very stable for three people to have their meals. Such a design fully utilizes the space of the aisle.

由于本案设计中空间是一个长方体，而且只有阳台一面是采光的，所以用衣柜来划分空间。但考虑到长方形的长度不够，所以扩展到阳台上，利用斜面来拉长空间并解决采光的问题。采光方面设计师在衣柜的中间与顶上，通过在通透的玻璃上贴钻石形贴膜的做法来达到效果，书桌旁边隔断采用钻石形采光口，从而充分解决客厅与厨房的光线问题。

厨房与客厅过道采用的是活动餐桌，不用餐时可以转到电视墙一面作为层板架摆放饰品，转到中间位置会有固定件固定，用餐时就比较稳固，还可供三人用餐，充分利用了过道空间。

# 公共・工程实景作品
## Public・Engineering Scene Works

项目名称：星艺装饰福建总部办公室　项目设计：广东星艺装饰集团　项目面积：400平方米　项目地址：福建福州　设计师：范林
项目用材：银白龙大理石、硅藻泥、灰镜、木通花

# 星艺装饰福建总部办公室
## Fujian Headquarter Office of Xingyi Decoration Company

本案设计定位于简约中式风格，黑色稳重的色块加上现代中式的木格，配合块面感强烈的前台，营造出一种新东方的时尚感。

基于公司"客户就是上帝，客户就是我们的服务中心"的营销理念，本案设计打破原有设计师前置的设计概念，大胆地设置了前厅、前台，再而进入中庭到设计洽谈区域，从客户进门到开始洽谈有个很好的过渡空间。

整体上看，空间不再具有浓烈的商业味，而是为客户营造了一种静心、舒适的氛围。

The style of this design is simple with Chinese characteristics. Chunks of black on the modern Chinese-style wooden framework match the reception area with strong outlines, providing a new sense of fashion featuring Eastern elements.

With the company's marketing philosophy that customers first and customers are the center of the service, this scenario breaks the design stereotypes of former designers. The designer boldly designs the antechamber and the reception area first, then the atrium and finally the negotiating area. Such an arrangement can provide a good transition for the customers from entering to the start of the negotiation itself.

Generally, the space of the office no longer radiates strong hint commerce, but rather, creates a tranquil and agreeable environment for negotiation.

项目名称："墨记"办公室　项目设计：广东星艺装饰集团　项目面积：230平方米　项目地址：广东广州　设计师：陈文辉　项目用材：生态木

# "墨记"办公室
"Ink Mark" Office

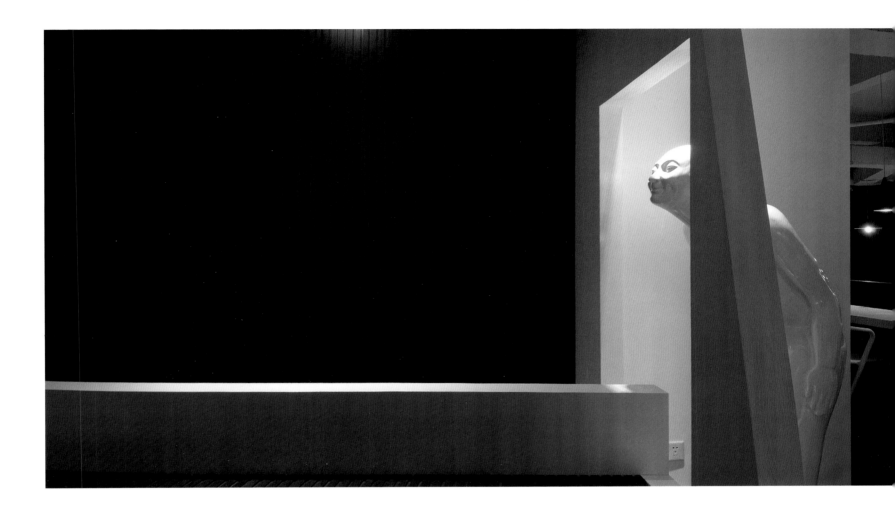

"墨记"是时间的记录,也是历史的记录。随着时间的推移,阳光的高度变化,光影也跟着缓慢推移,也记录着自然的变化,犹如《南方都市报》这个屹立在改革开放最前沿的广州媒体,它记录着改革开放的历史轨迹。墨象呈现表达着继承传统的责任,也承载着开拓进取的思维。光线在室内自由游移,引起空间与人之间一场灵动的对话,将人带进一个富有创意的世界。

"Ink Mark" is a record of time and history. With the time passing, the height of sunlight changes, so does the shadow the sunlight casts. This is also a record of the changes of nature, just as Nanfang Metropolis Daily, which is based in Guangzhou, the forefront of China's reform and opening up policy and the witness of China's development in this policy. The image of ink represents the responsibility of inheriting the traditions as well as the enterprising and aggressive mentality. Sunlight moves freely indoors, motivating a conversation between men and space and introducing people into a creative world.

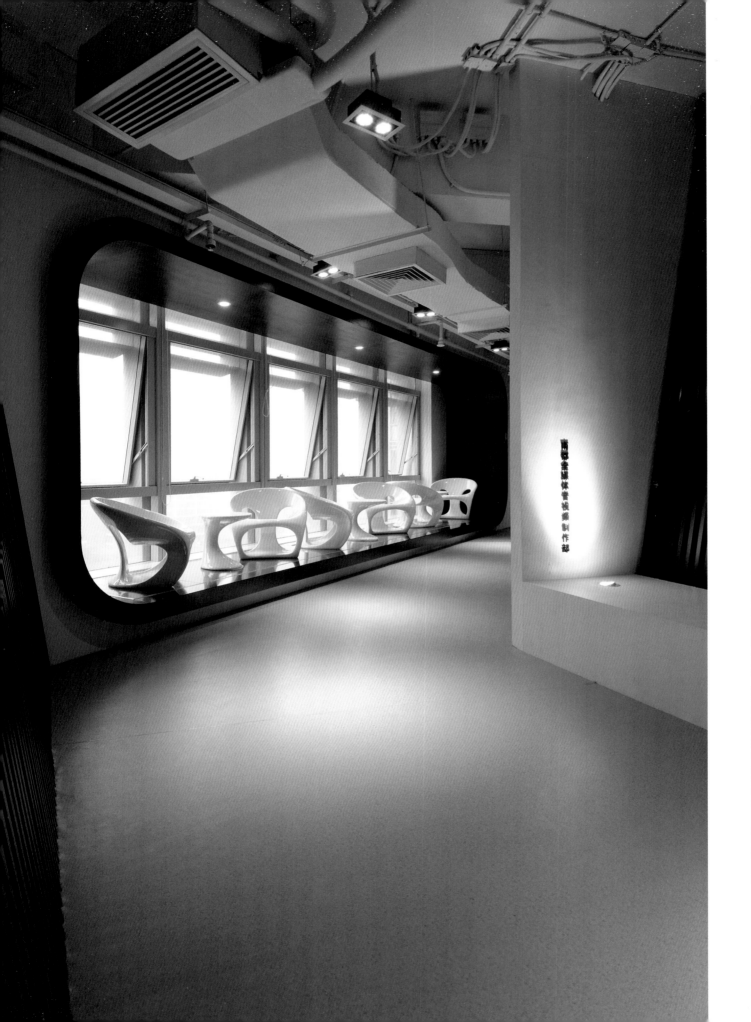

项目名称：广西贵港明悦大酒店　项目设计：广东星艺装饰集团　项目面积：10 000 平方米　项目地址：广西贵港　设计师：范建国　凌立成
项目用材：阿曼米黄大理石、黑金花大理石、红樱桃木饰面、香槟金不锈钢、地毯

# 广西贵港明悦大酒店
## Guigang Mingyue Hotel in Guangxi

设计初期，设计师在经过地理环境、位置、消费能力、建筑面积、功能划分等多方面的调查，以及与客户的多次沟通后，把酒店定位为达到四星级标准的高端商务酒店。本案采用现代简约的手法表现出具有东方韵味的文化产品，使入住的客人能够感受到高端大气又优雅舒适的艺术气息。

At the start of this project, the designer communicated with the client for several times concerning the geographic condition, positioning, consumption capability, covered area, function division and other issues. After that, the hotel was positioned as a four-star top business hotel. This project highlights images of Eastern culture with a simple and modern style, creating a noble and elegant atmosphere for the customers.

　　一进大堂，整个空间以米黄色暖基调为主，并点缀以长条金色不锈钢造型，宽敞明亮且极具吸引力，具有中式祥云寓意的大理石地面拼花，倒映着椭圆形的层叠水晶大吊灯，华丽而沉稳，时尚又大方。大堂前台深色的木饰面，配合蛛网大理石以及用抽象手法创作的装饰画，无不体现出整个酒店庄重沉稳又具有亲和力的灵魂思想，这也成为贯穿整个酒店主题的主线。

　　自助餐厅摒弃了传统的中式风格，以屏风隔断为区域划分，统一的深色木纹为主要装饰表现手法，搭配不规律拼贴的浅黄色墙砖，简单大方，使每一个用餐人员都能感到轻松自然。

　　客房，在满足数量的基础上变化房型，以满足除了商务为主的不同客人的需求。每间客房均配备四星级酒店必备的设施，简洁的深色木边框与墙纸、布艺以及定制的成品家具相呼应，极具淡雅的东方韵味；卫生间的设计更给客人以无限的想象空间，通透大方却又有其私密性。

Beige, a warm color, is the base color in the lobby. The lobby is decorated with long golden stainless structures which are large, bright and appealing. There are Chinese patterns of Xiangyun (lucky cloud) on the marble flooring, and the marbles reflect the oval cascading crystal chandeliers. Such an arrangement is gorgeous, solemn, decent and fashionable. The reception counter, covered with wooden veneer in dark color, matches marbles with patterns of spider nets and decorative abstract paintings. Such a design epitomizes the hotel's ideal of solemnity and hospitality, and also underlines the theme of the hotel.

The cafeteria is not designed in traditional Chinese style. Screen partitions divide the whole area into sections. Decorations with dark wooden pattern are the main feature, going with irregular collections of light yellow wall bricks. This looks simple but decent, making every customer eating in the cafeteria very comfortable and easy.

There are sufficient guest rooms and on this basis, the types of the rooms are varied. The hotel aims to meet the demands of various customers in addition to those mainly for business. Each room is equipped with facilities essential to four-star hotels. Brief and dark wooden framework echoes with the wallpapers, the linen and the customized furniture, full of elegant Eastern images. The bathrooms give customers unlimited room for imagination because they are opaque but preserve privacy.

项目名称：凤凰博瀚艺术馆　项目设计：广东星艺装饰集团　项目面积：1600平方米　项目地址：广东广州　设计师：帅伯尤　项目用材：大理石、紫檀木、黑玻、进口地塑、镜钢、方通

# 凤凰博瀚艺术馆
*Phoenix Bohan Gallery*

本项目的业主是一位酷爱收藏古董花瓶和字画的成功人士。因家中收藏颇多，故业主萌生出把自己过往的收藏品一并展示出来分享的想法。

在功能上，设计师一开始就对其馆所进行了详细简明的分区，以达到大方、实用的效果。在设计上，整个艺术馆进行了现代中式的归纳和精简，并融合了时尚元素，如天花铝通、地塑、银镜……突显了整个空间的精致和美观。另外，在灯光的布局上，设计师力求以点对点的方式让展品得到更好的发挥，并使展厅显得更为简约大方。

The owner of this property is a successful person especially keen on collecting ancient vase, paintings and calligraphy works. With a large collection, the owner has an idea of displaying what he has collected to others.

In terms of function, from the start, the designer has conducted a detailed classification of the space for broadening the view and for practical use. As to design itself, the whole museum has been unified and simplified in a modern Chinese way. At the same time, popular elements, including aluminum tube ceilings, plastic and silver mirror, make the whole space look especially exquisite and aesthetic. Additionally, on the arrangement of lighting, the designer tried to introduce the method of point to point to best illustrate the items exhibited and to make the whole space look spacious and simple.

项目名称：830 工作室　项目设计：广东星艺装饰集团　项目面积：35 平方米　项目地址：贵州贵阳　设计师：杨磊　项目用材：青石板、竹椅、青砖、鹅卵石

# 830 工作室
*Studio 830*

本着通过自己的设计，还原或者说更理想化地创造人们心灵深处的空间的想法，一个设计者的专属记忆空间诞生。如果说，好的设计能改善人们的生活，那么能穿越时间和空间的设计，会使人的内心得到安定和谐。

Out of the designer's idea of designing, restoring or ideally creating a space deeply rooted in people's heart, a space exclusively full of the designer's memories came into being. If a good design can improve people's life, projects that transcend time and space can help people feel at peace.

本案设计师是一名从粤西来到贵州生活的设计者，设计自己的办公室是本着回归儿时生活的情境和能给自己一个保留传统空间的初衷。一个设计师也许可以阅读无数的空间和设计风格，但是他心里真正属于自己的空间只有一个。在设计者看来，他的梦想空间一定是他出生和长大的地方——古镇、老街、学堂……那些深深影响了他一生的地方却只能停留在曾经的记忆里。

This project is made by a designer who comes from Western Guangdong and lives in Guizhou. The original purpose is to design an office based on his childhood memories and for him to preserve traditions as well. A designer may have witnessed countless spaces and design scenarios, but there is only one room that is owned in his heart. In the sight of the designer, his dream space must be based on the places where he was born and grew up—the old town, the ancient streets and the old school, and those places have an impact on his mind for his whole life, but only remain in his memory.

项目名称：国宾时光汇售楼部　项目设计：广东星艺装饰集团　项目面积：1 000 平方米　项目地址：四川成都　设计师：徐戈　项目用材：大理石、橡木面板、不锈钢

# 国宾时光汇售楼部
Guobin Time Sales Department

国宾时光汇位于成都土桥金牛金科北路（交警六分局旁），本案主要是为售楼部进行装修，不但注重精美的现代设计的细部，而且以独特的方式展示了"动"与"静"、"声"与"色"结合的环境，无论是外部空间还是内部空间，都体现了妙趣横生的创意。

"Guobin Time" is situated in Jinniujinke Road North, Tuqiao, Chengdu, next to the sixth local traffic police station. This design is for a real estate sales department. The project not only focuses on the exquisite modern design itself, but also displays a unique integration of "motion" and "stillness", "sound" and "color". Signs of interesting innovation can be seen in both the exterior and the interior.

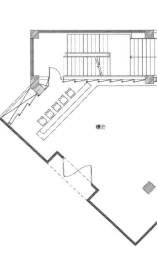

中间部位设置了沙盘展示区，而尾部区域设置了洽谈区，二者共享又不互相干扰，合理利用了空间。整个空间都以简约而不简单的思想来设计，每一处空间都有它独特的定义，所代表的含义也各有不同。设计如流水，在它温柔地敲击着我们内心的同时也在洗涤我们每个人的心灵。

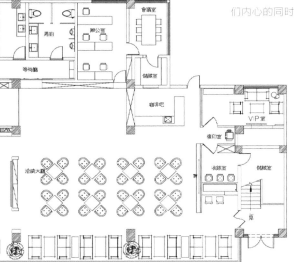

In the middle is a sand table of the real estate and at the end of the area, there is a section for negotiation. The two sections are linked but they do not cause disturbance to each other, utilizing the whole area properly. The whole space is decorated upon a concise and simple design, in which each section has its unique definition and meaning. The design is like streaming water, mildly knocking on our heart and cleaning every one's soul.

项目名称：光之游戏办公室　项目设计：广东星艺装饰集团　项目面积：300平方米　项目地址：江西上饶　设计师：陈文辉　项目用材：石材、砖

# 光之游戏办公室
## Game of Light Office

如果说实体创造了第一空间,光则创造了第二空间,而影的出现会创造出第三空间,最后,光与影的相互交织就形成了第四空间。

空间是承载光的容器,光线在室内空间游走、交织,就像谱写一篇光的乐章;本案设计中,设计师带领大家闭上眼睛,细心聆听光的声音和光的语言。

If the first space is created by entity, the second is by light, the third by the appearance of shadow and finally, the fourth space by the interaction of light and shadow.

Space is the container of light. It wanders and interweaves in the interior space as if composing a movement of itself. In this design, the designer may guide others to carefully listen to the sound and language of light with their eyes closed.

# 公共 · 方案设计作品
## Public · Scenario Design Works

项目名称：北戴河白公馆　项目设计：广东星艺装饰集团　项目面积：2 300 平方米　项目地址：河北秦皇岛　设计师：集团设计研究院
项目用材：实木、青石、玻璃、地毯

# 北戴河白公馆
## Mansion of Bai by Beidaihe

作为圈子文化的专属空间，本案仅服务于小众精英群体，并将风格取向定位为尊贵东方格调。作为有些年月的旧建筑，原本相对保守、低矮的格局不足以承载项目性质所赋予的使命，重新解剖和规划建筑骨架后，终于塑造出宽大高挑、修长挺拔的中央门庭，层次分明的主交通回廊贯穿其间。

整个空间格局高低错落有致，一扫之前低矮局促的形象。东方的贵族人文元素通过装饰手法、陈设品植入其间，强大而细致的配套功能恰当地设置在不同的楼层区域，移步换景，尽显专属的尊贵气息。

Exclusively for the culture of social circle, this project only serves a small circle of elites, with a distinguished Eastern style. This is an old building constructed for quite some time. Its original conservative and narrow arrangement of space can not live up to the requirement of the project. After the rearrangement and restructuring of the building, a grand central courtyard has finally been created with long and distinctive corridors.

The whole space is well-proportioned vertically, totally different from the previous narrow and constrained arrangement. Cultural element of Eastern aristocracy is shown through decoration and exhibited items. Specialized and useful facilities are properly positioned in different places. Various sceneries come from various perspectives, fully displaying the dignity of aristocracy.

项目名称：广西南宁波斯顿精品酒店　项目设计：广东星艺装饰集团　项目面积：11 000 平方米　项目地址：广西南宁　设计师：凌立成　许舰
项目用材：波斯灰大理石、地毯、木饰面、铜

# 广西南宁波斯顿精品酒店
## Nanning Bossdun Boutique Hotel in Guangxi

作为都市里的精品酒店，该案在设计之初希望酒店空间能够把人轻轻地包裹，使人得以安静地享受，在闹市凡尘中偷得半日清闲。

As an urban boutique hotel, it is originally designed to lightly envelope its customers, enabling them to enjoy its tranquility and leisure spared from the hustle and bustle of urban lives.

酒店迎宾大厅面积不大且结构比较曲折，但挑高6米，整个空间由多层次多质感的暖灰色构成，局部黑色镜钢点缀，纯净而优雅。这种看似单一，近乎固执的暖灰色应用一直贯穿始终，以一种简单而有力的节奏直指内心。虽然没有张扬出奇的造型，也没有浮夸闪亮的材料，但能让宾客在享受精品酒店的舒适之余，又能畅游在东西方文化交融的现代艺术氛围之中。

The hotel's reception hall does not cover a large area, but it is six meters in height. The whole space is covered by several distinctive warm hues of grey which are full of texture. Some places are decorated with black mirror frameworks, displaying purity and elegance. This seemingly monotonous and almost stubborn application of warm hues of grey has been insisted from the beginning to the very end. This application is empowered by a simple but strong beat, penetrating people's heart. Though without unusual appearances or glittering materials, the hotel can still offer a comfortable environment for the customers and a modern artistic atmosphere where they are involved in the integration of Eastern and Western cultures.

客房的设计摒弃奢华,追求优雅。看似单调的设计,却给予客人安静、轻松甚至宾至如归的感觉。客房面积不大却布置得精致灵活,设备齐全,执着的暖灰色设计在灯光下更显温馨。

Designs of the guest rooms focus on elegance instead of extravagance. Dull as it seems, such designs can bring peace and relaxation to customers and even make them feel at home. Rooms are not very large, but exquisitely designed with complete facilities. Therefore, the warm hues of grey appear even warmer in the lamplight.

项目名称：山水城会所　项目设计：广东星艺装饰集团　项目面积：3 600 平方米　项目地址：江西九江　设计师：集团设计研究院
项目用材：松木、灰木纹大理石、玻璃

# 山水城会所
## Landscape City Club

作为一栋依山傍水、轴线对称均衡的建筑，空间设计需要表现这些特征和优点。让整体场景流露出惬意、舒适的度假氛围是本设计的重点，经过规划，将宽大高挑的中央接待大厅通过巨型的木质坡屋顶和两翼延展开的宽阔接待区层次分明地连接起来，并通过艺术隔屏赋予空间层叠交错的视觉感，中央大厅采用唯一的一组巨型定制大灯悬浮在半空，空灵而不乏韵律。

As to a building situated at the foot of a hill and beside a river, those geographic features and advantages should be highlighted in design. The focus of this project is to create a pleasant and comfortable atmosphere. In accordance with the plan, the grand central reception hall is connected through the large wooden slope roof to the spacious reception zone with two flanks on its sides. Artistic partitions have made complicated divisions out of the whole space. In the central hall, there suspends a large collection of pendant lights in mid air. The lights seem to float in the air with rhythm.

项目名称：天瑞阳光酒店　项目设计：广东星艺装饰集团　项目面积：28 000 平方米　项目地址：云南香格里拉　设计师：熊卫明
项目用材：石材、实木、墙布

# 天瑞阳光酒店
Tianrui Sunshine Hotel

香格里拉天瑞阳光酒店地理位置特殊，它位于香格里拉县城西北部，紧靠214国道与中德路交汇处的南端600米处，地块现状与214国道道路高差有4米左右，北面为住宅规划用地，西面紧邻小型采石场，东临214国道，西南边为一座山坡。

Shangri-la Tianrui Sunshine Hotel has a special geographical location. It is situated in the northwest of Shangri-la County, 600 meters to the south of the convergence between 214 National Avenue and Zhongde Road. The hotel is about 4 meters higher than the 214 National Avenue. A land planned for residential construction is on the north of the hotel, a small-scale quarry on the west, 214 National Avenue on the east and a hill on the southwest.

Based on the fact that this hotel is the most typical in Shangri-la, the idea of design focuses on highlighting its dignified identity and its ethnical characteristics. The personalized style of design integrates with Tibetan design elements, with four major hues of red, white, blue and yellow to strengthen visual effect. This method sets apart the solemn but not monotonous characteristics of Tibetan design. Along with a large coverage of carpet, an elegant and comfortable space comes into people's eyes.

The roof of the hall is decorated with beams, typical of Tibetan style. The meeting room is illuminated with a combination of golden and white lights. Covered in carpet, the meeting room has been given a broad, bright and comfortable environment. Hidden elements of Tibetan style have been applied in designing the guest rooms. Simple and modern rooms are adorned with carpets with Xiangyun patterns and in four major hues of red, white, blue and yellow, which are unique in Tibetan style. Such arrangement creates a solemn and pleasant atmosphere in the rooms.

The aim of this design is to highlight the rich Tibetan culture, introducing the personalized and modern elements into the ancient Tibetan design at the same time. This can provide a new perspective for Tibetan design and help promote the cultural characteristics of minor ethnic groups.

鉴于该酒店在香格里拉最具代表性，于是设计的重点在于彰显尊贵身份的同时又突显民族特色：在个性化的现代风格中运用藏式设计元素，利用红、白、蓝、黄四个主色调来突显视觉效果，突出了藏式风格沉稳又不失单调的民族特色；配上大面积的地毯，塑造出优雅而舒适的空间。

大堂的整个顶部运用假梁彰显藏式风格。会议室则运用金色以及白色配上合理的灯光，加上大面积的地毯，营造出一个宽敞明亮且舒适的环境。酒店房间运用了隐藏式藏式设计元素，让简洁的现代风格房间通过藏式特有的红、白、蓝、黄四大主色调和祥云地毯来点缀，让房间整体氛围感觉沉稳而舒适。

设计主题强调香格里拉的藏式文化底蕴，在古老的藏式风格中融入个性化现代元素，给人一种新的藏式设计理念，延伸出现代少数民族特有的文化特色。

项目名称：车博汇会所  项目设计：广东星艺装饰集团  项目面积：6 000 平方米  项目地址：云南昆明  设计师：熊卫明
项目用材：石材、黑镜

# 车博汇会所
*Car Bo Club*

设计以突显尊贵和气度为首要目标，以迎合高端客户的要求。设计中，中式的沉稳与现代的个性相融合，古典的柔美与现代的硬朗相碰撞，内敛与灵动相辅相成，从而塑造出大气而优雅的会所空间。

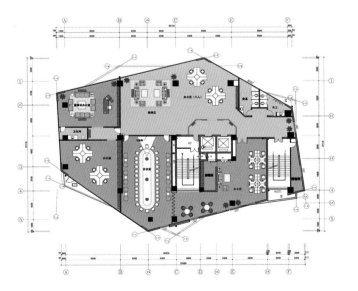

The priority of this project is to emphasize dignified identity and bearing to meet the demand of high end clients. In this project, the Chinese composure integrates with modern individuality; classical gentleness collides with modern hardness, while modesty and cleverness complement each other. In this way, a grandiose and graceful space is created in the club.

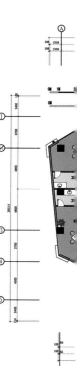

精美线条的视觉效果，配合高级的物料配饰；一楼以白色、黑色为主色调，有着鲜明的现代感，突出了项目特有的超前时代特性。二楼应用金色、米黄、暗红的宫廷系色调，有着鲜明的奢华感，突出了项目特有的高贵气派。餐厅中，镶着精美图案的屏风作为座位区的隔断，大量的灯具以祥云的切面形状出现，令简洁、古典中透露出一种透逸之感。三楼中个性的办公空间给工作提供灵动的氛围。四楼异域风格的SPA空间让人感知与世界的交融，缎面沙发靠垫、落地灯、墙面装饰一应俱全，而落地玻璃墙则将本案室外精美的园林景观予以尽情展现。

设计主题强调同一片天空下在不同的区域能创造出相同的品位，却延伸出不同的生活模式，给人另一种选择。气质自由、开放，处处显露着海纳百川的胸怀。

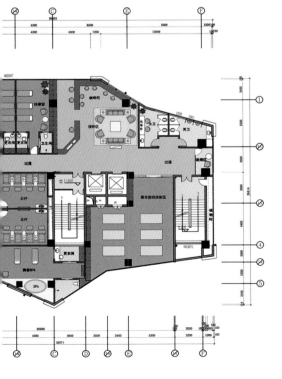

Visual effects of delicate lines go along with premium accessories. Black and white are basic hues for the first floor, with a distinct sense of modernity and a focus on the project's fashionable features. Gold, ivory and dark red, which are palace colors, are used for the second floor. Such colors are of distinct extravagance, outlining the specific dignity of this project. In the dinning hall, partitions screenings with delicate patterns divides the hall into sections, and a lot of lights appear in the form of Xiangyun. By such design, a sense of winding adds to the simple and classical scene. On the third floor, individualized offices offer a flexible and smart environment for work. On the fourth floor, the exotic spa room provides a place for the integration of people's conception and the world. Satin sofa cushions, lamplight and wall decorations are all in place and the glass wall provides a full view of the exquisite exterior landscape of this project.

For this project, the idea is to emphasize that although different zones can produce the same taste under the sky, various lifestyles are derived out of those zones, giving people alternative choices. Both freedom and openness represent the attitude of inclusiveness.

项目名称：桂林香樟楼大酒店　项目设计：广东星艺装饰集团　项目面积：9 000平方米　项目地址：广西桂林　设计师：范建国　许舰
项目用材：波斯灰大理石、杭灰大理石、核桃木饰面、山水主题墙纸、黑钛钢线条

# 桂林香樟林大酒店
*Camphor Forest Hotel in Guilin*

设计者通过对桂林作为国内一线旅游城市特色元素的提炼，营造出富含传统地域文化内涵的现代精品酒店空间。通过"形"与"色"的组合转化，突出"形色桂林"的主题概念。

Having refined the characteristics of Guilin as one of China's top tourist destinations, the designer has created a modern boutique hotel rich in the regions' traditional culture. With the integration of "shape" and "color", the concept of "Guilin's shape and color" is highlighted in this scenario.

从整体入手，从使用功能出发，实行以人为本原则，在建筑的基础上达到深化、细化，使建筑体与周边环境相融汇，一气呵成。

一流的建筑，一流的设计，并不等于昂贵材料的堆砌，更多的在于它的独特性、合理性和人文性。

本案的设计宗旨为：实用美观，命题规划，降低环境污染；以人为本，绿色环保，提高运用产能。

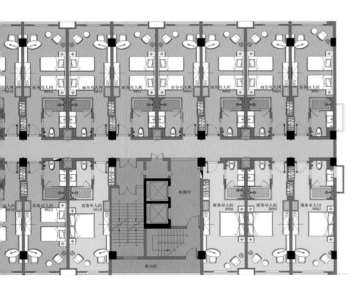

Overall, from the perspective of the hotel's function, customers are the top priority of this project, with profound and detailed division of functions in the building. As a result, the hotel can be involved in its surroundings.

A first rate building with an excellent design is not about putting together expensive materials, but rather about its uniqueness, rationality and significance in humanity.

The purpose of this project is to design a functional and aesthetic building which can help reduce pollution, protect the environment and improve energy efficiency.

项目名称：弘峰国际办公楼大堂　项目设计：广东星艺装饰集团　项目面积：8 000 平方米　项目地址：广东广州　设计师：集团设计研究院
项目用材：木、大理石、清水砖、水泥土坪漆

# 弘峰国际办公楼大堂
Lobby of Hongfeng International Office

创意以"服装、手提包"的明亮及时尚为设计表达的依据,同时加入创意的空间元素,朴实而又细腻。

This scenario is based on the brightness and fashion of "clothing and handbag" and at the same time, the spatial elements of innovation are added to this design, so that the hall looks simple but exquisite.

项目名称：愿佛宫殿展厅　项目设计：广东星艺装饰集团　项目面积：300平方米　项目地址：广东广州　设计师：集团设计研究院
项目用材：麻石材、木材、石头漆、佛像等

# 愿佛宫殿展厅
## Exhibition Hall of Wishful Buddha Palace

以佛殿的空间气氛为设计指导，取石窟和藏传佛教的宫殿形式，以现代的材料来表达，并突出打坐悟禅的冥想空间。结合经营佛教纪念品及佛具，加以商业灯光进行烘托。

The design is based on the atmosphere of Buddhist halls, and takes the form of grottos and palace of Tibetan Buddhism. Modern materials are used in construction and the space for meditation is also the focus of this scenario. The hall sells Buddhist souvenirs as well under commercial lighting.

项目名称：嘉莉诗国际旗舰店　项目设计：广东星艺装饰集团　项目面积：1300平方米　项目地址：广东佛山　设计师：集团设计研究院
项目用材：玻璃钢、玻璃、艺术马赛克、烤漆板

# 嘉莉诗国际旗舰店
## Flagship Store of Jealousy International

本案的原建筑特征和格局相对零碎，空间功能划分的手法相对陈旧。色彩运用杂乱而花哨，并且建筑外观没有彰显出作为本品牌的旗舰商业地产应有的夺目风采。在重新以草图的形式对整个空间内外部做分析的过程中，我们希望能以纯粹的女人的话题及元素来诠释一个女性的专属空间。一笔温柔润泽的弧线成为本案的主结构线。这条接近椭圆的结构线，勾勒出了旗舰店主前厅形体的同时，对整个建筑的外立面产生了极大的影响。

原本在建筑外立面开的几个展示橱窗，机械而保守，这笔弧线在建筑最显要的位置围合出了两幅巨大的橱窗，并呈扇形，以最大的广角视野充分地展现给世人。透过以编织的手法塑造出的整体建筑外立面的不规则网格看过去，能感受到含蓄而高贵的气质。通过草图的立体深化，这笔椭圆形的弧线最终过渡到一个拱起来的，类似于记忆中盛装女人结婚钻戒的那个柔软的丝绒面料的小盒子的形状，这也就是旗舰店主前厅的形状了，凹凸有致，设计的思路开始清晰而有趣起来。

The structure of the original building was not properly organized and the division of function in the building was also out-dated. Colors are gaudy and are used without any regular patterns. In addition, the building's exterior didn't underscore the prominence of the building it deserves as the flagship property of this brand. While analyzing the interior and exterior of the whole building with drafts, we hoped we could restore an exclusive space for women with female topics and elements only. The main structural line is a soft and smooth curve, almost the shape of an oval. It is the outline of the front lobby and also has a big impact of the building's exterior.

Several previous shop windows in the building's exterior were of technical and conservative design. However, two massive shop windows are outlined by the curve in the most prominent position of the building's exterior. Those windows take the shape of a fan, providing a maximum view for everyone. The irregular net shaped exterior with interwoven lines enables people to feel the modesty and nobleness of the brand. With the intensification of the draft's stereoscopic effect, the oval curve finally transforms into a shape of a small box with soft velvet lining which holds a woman's diamond wedding ring. This is the outline of the front lobby as well, with soft ups and downs of a curve, and the idea of design starts to become interesting and clear.

在前厅的中央，应用GRC异型材质做了一枚巨型的很怀旧很简约的戒指形状的中央陈设装置，并呈倾斜状置于正对主入口的前厅中央，正对主厅的圆形中空部分，视觉效果贯穿一、二楼。而从主入口方向看到的扣过来的那枚巨大的戒指，正是我们要追求的那种尊贵神秘、欲罢不能的视觉感受。盛放货品的展示架被环形安置于中央装置的周边，整个空间尽显时尚奢华。

In the middle of the front lobby, there is a large reminiscent and simple structure shaped as a ring. It is placed aslant and faces the main entrance of the lobby, vertically under the round hollow part of the lobby's ceiling. This structure can be seen on the first floor and second floor. The large ring activates the mysterious and unstoppable visual sensation among people. Shelves with commodities are placed around the central structure, so that the whole space looks fashionable and luxurious.

项目名称：清远狮子湖私人会所　项目设计：广东星艺装饰集团　项目面积：680 平方米　项目地址：广东清远　设计师：汪克成　罗东知　项目用材：艺术红砖、水泥、原木

# 清远狮子湖私人会所
Lion Lake Private Club in Qingyuan

设计师用有形的墙面对比红砖材质,明与暗、新与旧在空间中冲突又和谐并存,让会所这一传统空间呈现一种怀旧的时代感。裸露的素水泥材质和自然的木质顶面相呼应。品味酒文化发展的同时,让空间使用者获得平和恬淡的心理感受。

Red bricks are prominent against the tangible walls. Bright and dark, new and old contradict each other, but also exist in harmony, bringing a sense of nostalgia to this traditional space. Bare cement material echoes with the natural wooden roof. While exploring the development of the Chinese wine culture, users of this club can feel at peace with themselves.

细腻与粗犷之间的恰到好处,创造出一处相对私密舒适的环境。造型和环境糅合后的浑然天成,带给人一种熟悉的舒适感;结合酒窖空间本身的需求对光线进行二次控制,在满足各种功能的同时将环境和使用习惯紧密结合在一起。

Delicacy and roughness are at a proper balance, creating a relatively private and comfortable atmosphere. The natural integration of design and the environment offers a familiar sense of comfort. With the demand of the wine cellar taken into account, the lighting of the room is under double control. While all the functions are fulfilled, the environment and people's habits are closely connected to each other.

项目名称：长征电器办公楼大堂　项目设计：广东星艺装饰集团　项目面积：13 975 平方米　项目地址：贵州遵义　设计师：集团设计研究院
项目用材：鱼肚白大理石、法国木质灰大理石、本色不锈钢、黑钢橡木

# 长征电器办公楼大堂
## Lobby of Long March Electric Appliance Office

本案以外观及其国际化智慧写字楼的定位为设计基础要因，以秩序、齐整、组别式的空间语言，强调一种竖向、横向线条为造型母体，塑造一个有着现代、向上进取氛围的写字楼办公空间。竖向线条配合石材、金属的对比设计手法，使建筑物整体层次丰富，韵律感极强，同时智能化、高效率等理念也在设计中得以展示。

设计强调多种功能关系的同时，也注重空间品质与建筑气质的共生。由于此空间主体是办公室写字楼，设计师也关注在文化内涵上与主题相呼应。平面功能布置充分考虑了人流、管理安全、信息化、人性化等因素。

This project is oriented towards the exterior appearance of the building and its position as an international intelligent office building. Orderly and classified spatial language outlines a basic structure consisting of vertical and horizontal lines, while cultivating a modern and enterprising atmosphere in office. Vertical lines along with the contrast between stones and metals divide the whole space into various layers and bring about a strong sense of rhythm. At the same time, concepts of intelligence and high efficiency are shown in this project as well.

This scenario focuses on various functional relationships as well as the coexistence between the quality of space and the feature of the architecture. The major part of this project is the office building, and thus, the designer attaches importance to the interaction with the cultural connotation and the theme. Factors including human flow, administration and security, computerization and humanization are taken into account in functional arrangement.

项目名称：菲音游戏办公室　项目设计：广东星艺装饰集团　项目面积：10 000 平方米　项目地址：广东广州　设计师：集团设计研究院
项目用材：不锈钢、地坪漆、玉石

# 菲音游戏办公室
*Feiyin Game Studio*

以现代简约明快的色彩，结合游戏空间，创造舒适、个性的办公环境。

The objective is to create a personalized and comfortable office environment by adopting a video game environment and using bright and basic colors.

项目名称：柏川私人会所　项目设计：广东星艺装饰集团　项目面积：600 平方米　项目地址：广东广州　设计师：集团设计研究院　项目用材：灰石、木、白色漆

# 柏川私人会所
## Bochuan Private Club

打造中国风的私人小型会所，集造景手法，凸显品味中国居住的文化空间，宁静中透着禅意。

The scenario is to make a small private club featuring Chinese characteristics with methods of creating landscapes, experiencing the culture of Chinese residential spaces and serenity from those peaceful places.

项目名称：恒力城奢侈品店　项目设计：广东星艺装饰集团　项目面积：130平方米　项目地址：福建福州　设计师：范林　项目用材：PVC管、水泥砂浆

# 恒力城奢侈品店
## Luxury Store in Hengli City

这是由一小区套房改造而成的一个奢侈品店,业主要求简洁且富有强烈的时尚感,但又不一定需要花费较大的价格去装修,并希望能体现奢侈展品的档次。

本案空间设计简洁,又不乏庄重和凝练、舒展和大气,以灰黑色来衬托奢侈品的尊贵。

墙面的造型用 PVC 管切半拼接,再喷涂厚重的水泥砂浆做粗糙的肌理效果,从而有效地衬托出精致的商品。客户走在黑色的设计空间里,也能在琳琅满目的奢侈品中尽情享受其带来的高贵体验。

The shop of luxury goods is reconstructed from an apartment. The owner of the property required that the design be concise and very trendy, but not necessarily with a huge investment of money into decoration. The owner also hoped that such a design could display how exquisite those luxury items are.

The design of this space is simple, but not without solemnity, precision, extension and majesty. Grey and black can reveal the nobleness of the luxuries.

The walls are decorated by joining PVC pipes which are cut in half, and spraying thick layers of cement mortar to produce a coarse texture. This can provide a good background for showing delicate commodities. Walking in the black space, the customers are on a noble experience brought by a large collection of luxuries.

项目名称：南方都市报业传媒办公室　项目设计：广东星艺装饰集团　项目面积：700平方米　项目地址：广东广州　设计师：陈文辉　项目用材：石材、开石

# 南方都市报业传媒办公室
## Office of Southern Metropolis Daily Media

本案是南方报业集团的全媒体集群办公空间。全媒体集群是把传统媒体与新型媒体完美结合起来,形成一个全新的媒体概念。

This project concerns the office area of Southern Media Group. The media group is a perfect combination of traditional media and new media, shaping a brand new concept of media.

入口墙面利用石材原有质感及开石形成的天然肌理，形成开山劈石的效果，从视觉上强调南都全媒体开拓、进取的精神；两边石材分离开后展现出来的是南都全媒体的传统媒介——报纸，利用报纸叠加形成的切面带来视觉上的冲击；显示屏下身着汉服的简笔人物，让人处处感受到对传统媒介的坚持和对新型媒介的追求。

本案设计关键词是"开拓、进取、坚持、包容"。27个现代名家设计的不同座椅，体现了27个人不同的思想，亦体现了南都全媒体集群对不同思想的包容。

The texture and the natural pattern of the stones on the entrance wall create an image of extracting stones out of mountains. This visual effect emphasizes the enterprising and pioneering spirit of ND Media. After the separation of the stones on both sides, the traditional medium of ND Media—newspaper is revealed. The visual effect is sharpened with the image of newspapers stacking up. The matchstick figure in Han costume on the screen enables everyone to notice the persistence of old media and the pursuit for new media.

The key words of this scenario are "Exploration, Progress, Persistence and Tolerance". Seats designed by 27 famous modern designers represent 27 different ideas, and also reflect ND Media's tolerance of various thoughts.

耕艺种德

设计幸福